The Emerging Technology Landscape

Bashir Mir, PhD

Copyright © by Bashir Mir 2012

All rights reserved. This book, or parts thereof, may not be reproduced in any form without permission.

Library of Congress Cataloging-in-Publication Data

Mir, Bashir.

The Emerging Technology Landscape/ by Bashir Mir.

p.   cm.

ISBN 13: 978-1479286867

ISBN 10: 1479286869

Printed in the United States of America

To my wonderful family, for their continued love and support.

CONTENTS

Introduction..1

SECTION 1: SMART PRODUCTS

    Smart Blood Test..................................5

    Smart Windows....................................7

    Smart Water Supply..............................9

    Smart Headlight.....................................11

    Smart Hand Pump...............................13

SECTION 2: CLEAN ENERGY

    'Paint on' Batteries...............................17

    People Power......................................19

    Wastewater Generator........................23

    Biogas..25

    Artificial Leaf.......................................27

SECTION 3: ECO-FRIENDLY PRODUCTS

Next Super Material……….…………………....…31

Golden Fiber……………………………………...33

Spider Goat……………………….......................35

Green Manufacturing………..……………….…...37

Spider Silk Electronics…………………………...39

Eco-Ink………………………………………....41

Self-Cleaning Fabric………………………..…43

SECTION 4: COOL TECHNOLOGY

Touch-Less Surgery…………………………..47

Ultrasound for Fractures………..………….…...49

Self-Driving Car……………………....................51

Self-Driving Convoy…………….…………….53

Robotic Fish……………………………….....55

Mine Kafon………………………………….57

Voice-Controlled Television……………………...59

Virtual Grocery……………………………..…..61

Mirage Table……………………………………..63

Ultra-Thin Glass…………...………………….....65

Virtual Changing Room…..…………………....67

Google Fiber………………………………….....69

Everyday Objects…………………………….....71

Introduction

Innovation can take many forms. It can emerge in the form of incremental improvements to existing products, the creation of breakthroughs that result in entirely new products and services, cost reductions, efficiency improvements, new business models, and new ventures. In the quest for innovation, very few ideas that emerge at the input stage translate into useful innovations at the output stage. In other words, one can visualize the innovation process as a funnel: with a plethora of ideas at the wide end, and a few marketable innovations at the narrow end of the funnel.

During the last two years, I started collecting stories of innovation. In this journey, I came across both emerging technologies that are shaping our future and old innovations that are becoming more relevant during the recession. I also observed a renewed focus on environmentally friendly products. This book is based on these stories and covers 30 innovations from diverse areas that are likely to have significant impact on our lives. These innovations are divided into four sections. Section one covers smart products and describes innovations such as smart blood test, smart windows, smart water supply, smart headlight, and smart hand pump. Section two focuses on clean energy and describes innovative solutions to generate and use clean energy. The products in this category are artificial leaf, 'paint on' batteries, and power generation from humans, waste water and biogas. Section three describes seven eco-friendly products that combine both conventional and new technologies. In this category, the conventional products with a new twist are bamboo and jute. The new products are biotechnology based manufacturing of spider silk and other products, spider silk electronics, eco-ink, and self-cleaning fabric. Section four centers on technologies that will make our lives "cool" and comfortable in myriad ways. The technologies in this category range from everyday mundane objects to robotics to new medical technologies. The list includes touch-less surgery, ultrasound for fractures, self-driving car, self-driving convoy, robotic fish, mine kafon, voice controlled television, virtual grocery, mirage table, ultra-thin glass, virtual changing room, and google fiber.

SECTION 1

SMART PRODUCTS

Smart Blood Test

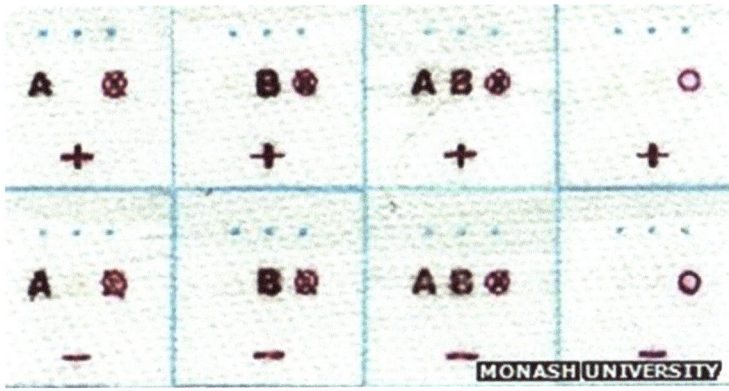

The paper makes blood type letters appear - written in blood

A team from Monash University in Australia has developed a paper-based sensor that writes blood type as text. The sensor may help non-experts interpret results rapidly. At times, people conducting blood tests at home, or even specialists in developing regions, make mistakes while interpreting a blood type test. These mistakes could have grave consequences. But with a device that can spell out the patient's blood type in written text, people can easily determine their blood type. Places in which the strips might be used include rapid response scenarios, such as battlefield casualties and automobile accidents.

When they compared the sensor's performance with mainstream blood typing technologies used in hospitals and pathological laboratories around the world, the Monash University team found that the sensor has the same accuracy—but is cheaper, faster, and easier to use. Even though the techniques behind the test are the same as conventional methods, the major novelty is in the ease of reading the strip, which spells out the letters for specific blood types. The device consists of a sensor made from a tiny piece of paper, coated with a hydrophobic (water-repellent) layer, with four "windows" without the water-resistant layer, making them prone to absorb liquid. Each area is shaped differently; for instance, one has the shape of the letter A, another the shape of the letter B. These areas are filled with antibodies that interact with red blood cells, making them clump together, or agglutinate, depending on the blood type. When a drop of type A blood fills the area of the paper containing antibodies corresponding to that type, the red blood cells form clumps and get stuck in the paper fibers, making a letter visible. The result remains even when the sensor is rinsed. AB type gives a red tint to both A and B-shaped windows. Since type O has no antigens and thus does not interact with any antibodies, the researchers shaped the third window as the letter X and filled it with antibodies against A and B. They then printed a letter O in the window with red waterproof ink. Blood types A, B, and AB made the X red, eliminating the O type by literally "crossing" it out. But if the sample is type O, the X becomes white after rinsing it with saline solution, and the red letter O remains.

Smart Windows

'Smart' window switches to dark mode to save energy

South Korean scientists have created a new type of "smart" window that switches from summer to winter mode. The window darkens when the outside air temperatures soar and becomes transparent when it gets cold in order to capture heat from the sun. Although "dimming" windows already exist, it is often necessary to switch them from winter to summer mode and back manually, using additional equipment such as home-automation panels. Smart windows allow for an almost instantaneous switch from opaque to transparent. This type of light control system might help households save on heating, cooling, and lighting costs by managing the light transmitted into the interior of a house. Smart windows can prevent the inside of a building from becoming overheated by reflecting away a large fraction of the incident sunlight in summer. Alternatively, they can help keep a room warm by absorbing the sun's heat in winter.

The existing technology uses charged particles called ions sandwiched between panes of glass. Electric current is then applied to switch the window from opaque to clear and back. But Ho Sun Lim, of Korea Electronics Technology Institute, and Jeong Ho Cho and Jooyong Kim, of Soongsil University, decided on a different approach. They used a special polymer, charged particles known as counter-ions, and solvents, such as methanol. The result was a glass that was much cheaper to manufacture and much less toxic than those currently available on the market. The window is able to switch from 100 percent opaque to almost completely clear in a matter of seconds.

Smart Water Supply

Israeli start-up Takadu helps Londoners save water

As water gushes through the labyrinthine infrastructure of the London water supply system, an ageing pipe creaks, whines noisily and finally bursts. Within seconds, an alert starts flashing on a remote computer in the tiny office of Takadu, an Israeli start-up in Tel Aviv. The information is then transmitted to Thames Water, the utility company responsible for bringing water to Londoners. Takadu is one of many companies around the world that aim to help save water by monitoring infrastructure and detecting leaks in pipes early.

According to the World Health Organization, about three billion people on Earth—almost one in two—live in water-scarce conditions, with demand growing drastically while supply remains constant. Out of all the water that's being supplied to consumers, more than 45 billion liters per day globally are lost to leakage—around 20 to 30 percent in developed countries, and close to 50 percent in developing ones. For instance, in the UK, it is estimated that around 3.3 billion liters of water are wasted every day, mainly in pipes. That's the equivalent of 1,000 Olympic swimming pools. Utility companies as well as consumers have to pay the price. Leakage costs the world's water supply firms approximately $14 billion per year, according to the World Bank. To keep the consumer "watered" enough, it is estimated that a global total of $23 trillion will be spent on improving public infrastructure that handles water and sewage from 2005 to 2030. To reduce operational costs, utility companies are turning to innovative solutions and new technologies to detect leaks early and eliminate them as soon as possible.

Just a few years ago, water supply firms in the UK used a number of devices to detect leaks, including "listening sticks," detectors that workers put to their ear to listen for unusual underground leak sounds. But with the rapid penetration of the internet into the world of sensors, everything has started to change. This is where Takadu comes in. It's a 24/7 computer watchdog. Takadu operates in big metropolitan areas, picking up data from different network meters, such as flow, pressure, and others. If the data shows that something is wrong—a small leak, a big burst, faulty equipment, or just a technician who left a valve open—they determine the location, the magnitude, and when it started, and then send the data straight to the repair team.

In the past few years, this "smart planet" concept has been leading the agenda of many companies worldwide. Initiatives ranging from smart meters and smart grids in buildings that aim to save energy, to smart parking places, intelligent cars, and even smart operating systems for entire cities give us a glimpse of what our world might be like in the very near future. Water technology developers are not lagging behind. Takadu is not the only company that uses smart water technology, but this tiny start-up is one of the market leaders. The firm was established with the goal of supplying enough water to Israel, which is located in a region where everyone is constantly aware of water scarcity. But the internet allowed it to work with countries all over the world.

Smart Headlight

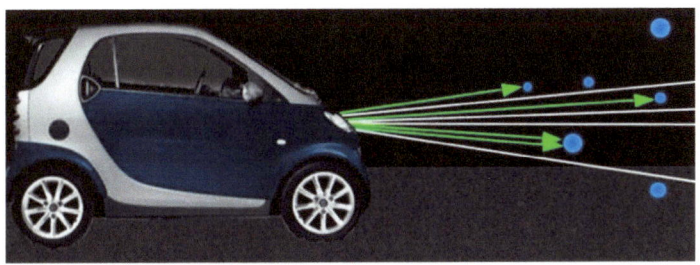

The system switches off light rays that hit the raindrops

Researchers in the US have come up with a solution to the problem of headlight glare reducing driver visibility in the rain. Scientists at Carnegie Mellon University have developed a smart headlight that can shine "around" rain. The idea is that the headlight will be able to predict where rain falls and adjust light beams accordingly.

Using low-cost, off-the-shelf components, the researchers set about developing a system that switches off rays of light that hit raindrops. The smart headlight consists of a projector, camera, and beam-splitter. The camera takes images of the raindrops; a processor uses a predictive algorithm to work out where rain will fall. The projector switches off light rays that would have normally hit the raindrops. The process from capture to reaction takes about 13 milliseconds. The result is a slightly dimmer headlight, but one that blocks out glare from falling rain and snow. Testing of an early prototype in a lab using artificial rain found that the system worked better at slower speeds. The researchers simulated different car speeds and rainfall intensity and found that in severe thunderstorm rain, the system had a 79 percent success rate in making raindrops invisible when the car travelled at 30 km/h, while at 100 km/h that fell to around 20 percent. Therefore, the system needs improvement before it could be used in real cars.

Smart Hand Pump

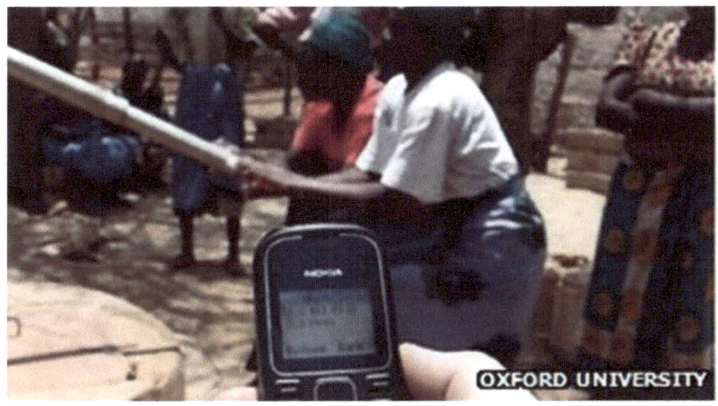

Mobile phone + hand pump = Smart hand pump

Rural communities across Africa may soon benefit from improved water supplies thanks to mobile phone technology. Hundreds of millions of people across rural Africa depend on hand pumps for their water supplies. But it is estimated that around one-third of the pumps are broken at any given moment. They are often located in remote areas, and repairs can sometimes take up to a month. But one of the big changes in Africa in recent years has been the expansion of mobile phone networks. It is now estimated that more people in Sub-Saharan Africa have access to these networks than to improved water supply.

Researchers at Oxford University have developed the idea of using the availability of mobile networks to signal when hand pumps are no longer working. They have built and tested implanting a mobile data transmitter into the handle of a pump. It measures the movement of the handle, and that data is used to estimate the water flow of that particular hand pump. It can periodically send information by text message back to a central office, where the data can be viewed. When a pump breaks, a mechanic can quickly be dispatched to fix it.

In the first phase, 70 villages in the Kyuso district in Kenya had smart hand pumps installed. The trial, which is funded by the UK's Department for International Development, will determine if the new system can cut the time taken to repair pumps. A number of big challenges remain, including the critical issue of power. The Kenya experiment will use long-lasting batteries, but the research team hopes to develop more sustainable ways of powering the transmitters. Another challenge is the threat of theft and vandalism.

Speedy repairs to hand pumps could have dramatic effects on local communities. Broken pumps mean more hours spent gathering the precious resource, and people often turn to unsafe sources of drinking water. But keeping the pumps working has an impact beyond delivering drinking water. Water does not just save lives in the short term—it is also a cornerstone for delivering economic growth and helping countries work their way out of poverty. The technology has other potential benefits. It will allow scientists to compile a real-time database of how much water is being used across the continent. Greater predictability of breakdowns could also help drive down the cost of repairs.

SECTION 2

CLEAN ENERGY

'Paint-on' Batteries

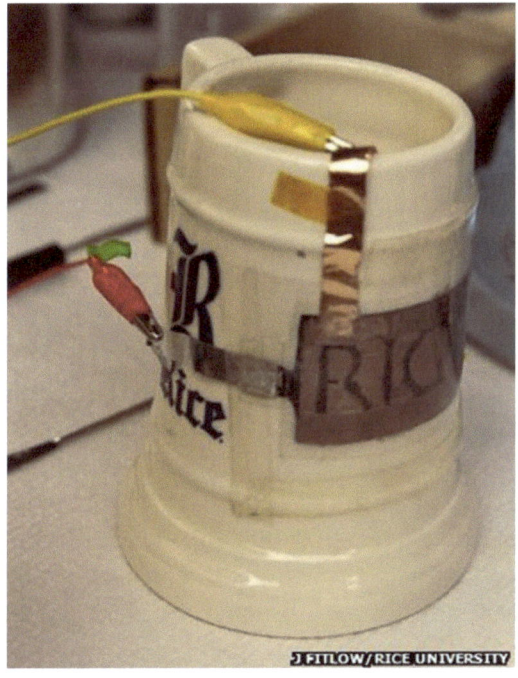

Batteries onto a beer stein

Researchers from Rice University in Texas have developed a means to spray-paint batteries onto any surface. Their batteries are made up of five separate layers, each with its own recipe, together measuring just 0.5mm thick. To demonstrate the technique, the research team painted batteries onto steel, glass, ceramic tile, and even a beer stein. The approach will be of particular interest in industrial applications, as it is compatible with existing spray-painting technology.

The most common batteries are made up of negative and positive halves (the anode and the cathode), a material to separate them, and current collector layers at the top and bottom to gather up the electric charges moving through. Many batteries are made in a kind of "Swiss roll" geometry, in which the layers are rolled up into a cylindrical or round-edged rectangular shape. But as more consumer technology is developed in challenging shapes and sizes, or "form factors," the need for batteries in non-standard shapes is rising. Flexible paper batteries have been developed, and there is clear interest in "structural batteries" built, for example, into the surface of electric vehicles. The new work opens up completely new avenues for putting batteries on nearly any surface in a simple and robust way.

Pulickel Ajayan and his colleagues chemically optimized the recipe for each of their five layers, using blends of chemicals common in lithium-ion batteries as well as novel materials, including carbon nanotubes—tiny "straws" of carbon with incredible electronic properties. But for the process to result in a working battery, all five layers must stick together and work in synchrony, and the tricky step was finding a separator material that kept the whole stack in one piece. When the team hit on using a chemical called poly-methylmethacrylate, they had a structure that would stick even to curved surfaces. This means traditional packaging for batteries has given way to a much more flexible approach that allows all kinds of new design and integration possibilities for storage devices. There has been a lot of interest in recent times in creating power sources with an improved form factor, and this is a big step forward in that direction.

People Power

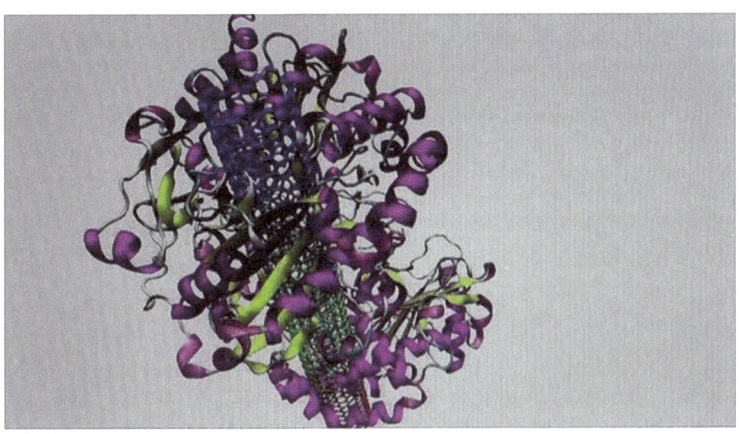

The cell uses glucose and oxygen to generate electricity

Plugging gadgets into a socket in the wall or loading them with batteries—or maybe even unfurling a solar panel—is how most of us think of getting electricity. But what about plugging them into your body? It may sound farfetched, but under the shadow of the Alps, Dr. Serge Cosnier and his team at the Joseph Fourier University in Grenoble have built a device to do just that. Their gadget, called a biofuel cell, uses glucose and oxygen at concentrations found in the body to generate electricity. They are the first group in the world to demonstrate a working device implanted in a living animal. If all goes according to plan, within a decade or two, biofuel cells may be used to power a range of medical implants, from sensors and drug delivery devices to entire artificial organs. All you'll need to do to power them up is eat a candy bar or drink a soda.

Biofuel cells could kick-start a revolution in artificial organs and prosthetics that would transform tens of thousands of lives every year. A new range of artificial, electrically powered organs are now under development, including hearts, kidneys, and bladder sphincters, and work has begun on fully functioning artificial limbs such as hands, fingers, and even eyes. But they all have one Achilles heel: They need electricity to run. Batteries are good enough for implants that don't need much power, but they run out fast, and when it comes to implants, that is more than just an inconvenience; it is a fundamental limitation. Even devices that do not use much power, such as pacemakers, have a fixed lifespan because they rely on batteries. They usually need their power packs replaced five years after implantation. One study in the US found that one in five 70-year-olds implanted with a pacemaker survived for another 20 years—meaning that group needed around three additional operations after the initial implant, just to replace the battery. Each operation is accompanied by the risk of surgical complications, something no one should have to face if it is avoidable. Other devices, such as artificial kidneys, limbs, or eyes, would have such high energy demands that users would have to change their power source every few weeks to keep them working. It is simply impractical to use batteries in these devices. This is where biofuel cells come in. Dr Cosnier and his team are one of a growing number of researchers around the world developing the technology in an attempt to sidestep this inherent limitation.

At heart, biofuel cells are incredibly simple. They are made of two special electrodes. One has the ability to remove electrons from glucose, while the other can donate electrons to molecules of oxygen and hydrogen, which produces water. Pop these electrodes into a solution containing glucose and oxygen, and one will start to rip electrons off the glucose, while the other will start dumping electrons onto oxygen. Connect the electrodes to a circuit and they produce a net flow of electrons from one electrode to the other via the circuit, resulting in an electrical current. Glucose and oxygen are both freely available in the human body, so hypothetically a biofuel cell could keep working indefinitely. A battery consumes the energy stored in it, and when it's finished, it can't be replenished. A biofuel cell in theory can work without limits because it consumes substances that come from physiological fluids, and those are constantly replenished.

In 2010, the researchers tested their fuel cell in a rat for 40 days and reported that it worked flawlessly, producing a steady electrical current throughout with no noticeable side effects on the rat's behavior or physiology. The electrodes were made by compressing a paste of carbon nanotubes mixed with glucose oxidase for one electrode, and glucose and polyphenol oxidase for the other. The electrodes have a platinum wire inserted in them to carry the current to the circuit. Then the electrodes are wrapped in a special material that prevents any nanotubes or enzymes from escaping into the body. Finally, the whole package is wrapped in a mesh that protects the electrodes from the body's immune system but allows the flow of glucose and oxygen to the electrodes. The whole package is then implanted in the rat. Implantation in a rat was a good proof of concept, but it had drawbacks. Rats are so small that the production of energy is insufficient to power a conventional device. Next, Dr. Cosnier plans to scale up his fuel cell and implant it in a cow.

Wastewater Generator

The prototype generates electricity from wastewater

Researchers at Pennsylvania State University have built a prototype device that can generate electricity from wastewater and simultaneously treat the water. According to the research team, the process could be used in developing countries to provide clean water and power.

Scientists around the world have been exploring the idea of generating renewable power along the coastline, where fresh water from rivers meets the salt water of the sea. Using a process called reverse electro-dialysis (RED), fresh water and sea water are placed in intermittent chambers separated by membranes, creating an electrochemical charge. The Penn State team says RED technology is problematic because of the large number of membranes required and because power plants have to be located by the sea. They claim the number of membranes can be reduced and the power output boosted by combining the technology with what are called microbial fuel cells (MFCs).These use organic matter in a solution—in this instance, wastewater— to create an electric current. The prototype technology also bypasses the need for salt water by using ammonium bicarbonate solution as a substitute, meaning the system could work in communities far from the sea. The ammonium bicarbonate solution would be constantly recycled, using waste heat from local industry. The main application right now is in wastewater treatment, in which you could effectively treat the water and gain some extra energy from waste heat.

Biogas

Biogas for cooking food

Having emerged as one of the pioneers in producing biogas from cow dung, the Himalayan nation of Nepal is now successfully transferring its technical expertise to other countries. Several Nepalese experts have been travelling to countries in South East Asia and Africa to introduce the "clean," homegrown technology that helps save forests and reduce carbon emissions from fossil fuels. Biogas from cow dung is mainly used for cooking in rural areas and also for lighting in houses. Renowned for displacing smoky ovens with clean cooking stoves, the Nepalese model of biogas has won the prestigious Ashden award.

The Biogas Partnership Project Nepal, a collaboration between the government, donors, and non-governmental organizations, has already installed plants for nearly 300,000 households across the country. The project helps to reduce 7.4 tons of greenhouse gases per household per year and protects 250,000 trees during the same period of time. The expertise gained over the years has benefited many communities in several developing and less- developed countries, including Cambodia, Laos, Vietnam, Indonesia, Bangladesh, Bhutan, and others in Asia, as well as around 10 countries in Africa. The success in Vietnam in particular has been outstanding; more than 100,000 biogas plants have been installed there. Installations are growing in Indonesia, where the technology is referred to as "The Nepal model." In several African countries, Nepalese experts are not only helping the communities install biogas plants; they are also conducting training in schools.

The technology is simple and natural. Bacteria from the cow's stomach break down the dung and other waste in an underground airtight digester. In the absence of oxygen, the mixing of cow dung with water leads to a reaction that produces a gas comprising up to 70 percent methane. The digested slurry flows into an outlet tank and ends up in the compost pit. The gas is tapped from the top of the dome with a pipe that ends at the burner on a kitchen stove.

Artificial Leaf

Solar cell bonded to a catalyst can harness the solar energy

Researchers led by MIT professor Daniel Nocera have produced what they're calling an "artificial leaf." The artificial leaf—a silicon solar cell with different catalytic materials bonded onto its two sides—needs no external wires or control circuits to operate. When placed in a container of water and exposed to sunlight, it quickly begins to generate streams of bubbles—oxygen bubbles from one side and hydrogen bubbles from the other. If placed in a container with a barrier to separate the two sides, the two streams of bubbles can be collected and stored, and used later to deliver power. For example, this can be accomplished by feeding them into a fuel cell that combines them once again in water while delivering an electric current.

The device is made entirely of abundant, inexpensive materials—mostly silicon, cobalt, and nickel—and works in ordinary water. Other attempts to produce devices that could use sunlight to split water have relied on corrosive solutions or on relatively rare and expensive materials, such as platinum. The artificial leaf is a thin sheet of semiconducting silicon—the material most solar cells are made of—which turns energy from the sun into a flow of wireless electricity within the sheet. Bound onto the silicon is a layer of a cobalt-based catalyst, which releases oxygen. The other side of the silicon sheet is coated with a layer of a nickel-molybdenum-zinc alloy, which releases hydrogen from the water molecules. The new device is not yet ready for commercial production, since systems to collect, store, and use the gases need to be developed.

SECTION 3

ECO-FRIENDLY PRODUCTS

Next Super Material

Booming bamboo: The next super-material!

Bamboo's image is undergoing a transformation. Some now call it "the timber of the 21st Century." Today you can buy a pair of bamboo socks or use it as a fully load-bearing structural beam in your house, and it is said that there are some 1,500 uses for it in between. There is a rapidly growing recognition of the ways in which bamboo can serve us as consumers and also help to save the planet from the effects of climate change because of its unrivalled capacity to capture carbon. New technologies and ways of industrially processing bamboo have made a big difference, enabling it to begin to compete effectively with wood products for Western markets. It is estimated that the world bamboo market stands at around $10 billion today, and the World Bamboo Organization says it could double in five years. The developing world is now embracing this potential growth. In eastern Nicaragua, bamboo was until recently regarded by most of the local population as valueless, more as a nuisance to be cleared than a boon to them and their region. But on land that was once under dense forest cover, then turned over to slash-and-burn agriculture and ranching, new bamboo plantations are rising.

Golden Fiber

Bangladesh's 'golden fiber' comes back from the brink

When use of synthetics such as polythene bags became widespread, the demand for jute declined and many jute mills in countries like Bangladesh were shut down. Thousands lost their jobs and farmers shifted from jute to more profitable rice cultivation. With growing environmental awareness, jute, which is biodegradable, has become the preferred alternative to synthetic bags and has made a spectacular comeback. Today, as demand increases, more farmers are returning to this traditional crop. It is estimated that nearly five million farmers are involved in jute plant cultivation in Bangladesh. It plays a key supportive role to the rural economy of Bangladesh.

Once the jute plants are harvested they are bundled together and immersed in running water and allowed to rot. Once it has become soft, the jute fiber is separated by hand. Then the fibers are stripped from the plant. The stripped fiber is dried and later sent to mills for processing. Jute is also versatile, strong, and long-lasting, and scientists say they are discovering more uses for it in different sectors. For example, Geotextiles, a diversified jute product, is used for soil erosion control and in laying roads to increase durability. By processing the fiber mechanically and by treating it chemically, jute can be used to make bags, carpets, textiles, and even insulation material. The natural fiber is also used to make pulp and paper. Bangladeshi scientists are now working on an ambitious project to blend jute fabric with cotton to produce denim fabric. They say if the jute plant is harvested earlier than the usual period of 120 days, then it gives a softer fabric.

Spider Goat

Transgenic goats produce spider silk protein in their milk

Researchers from the University of Wyoming and other universities have developed a way to incorporate the spider's silk-spinning genes into goats, allowing the researchers to harvest the silk protein from the goat's milk for a variety of applications. For instance, due to its strength and elasticity, spider silk fiber could have several medical uses, such as making eye sutures and artificial ligaments and tendons, and for jaw repair. The silk could also have applications in bulletproof vests and improved car airbags.

Normally, getting enough spider silk for these applications requires a large numbers of spiders. However, spiders tend to be territorial, so when the researchers tried to set up spider farms, the spiders killed each other. To solve this problem, scientists decided to put the spiders' dragline silk gene into goats in such a way that the goats would only make the protein in their milk. Like any other genetic factor, only a certain percentage of the goats end up with the gene. When these transgenic goats give birth and start lactating, the researchers collect the milk and purify the spider silk protein, thereby producing much higher quantities of spider silk. Other than their ability to produce the spider silk protein, the goats do not seem to have any other differences in health, appearance, or behavior compared to goats without the gene.

Green Manufacturing

Abalone, a marine snail, model for green manufacturing!

The humble abalone, a slimy sea snail that occasionally ends up as someone's dinner, pulls calcium and carbon from sea water and transforms these nutrients into a durable, protective shell. Crusty and dingy on one side, shimmery and alluring on the other, this body armor is 3,000 times stronger than chalk, which is its chemical equivalent. Abalone shells have inspired Professor Angela Belcher's work at the Massachusetts Institute of Technology for more than two decades and brought her to the pinnacle of science. And they have implications for the future of manufacturing, green energy, medicine, and science. Belcher's work unites the inanimate world of simple chemicals with proteins made by living creatures, a mash-up of the living and the lifeless. Like the abalone collecting its materials in shallow water and then laying them down like bricks in a wall, Belcher takes basic chemical elements from the natural world: carbon, calcium, silicon, and zinc. Then she mixes them with simple, harmless viruses whose genes have been reprogrammed to promote random variations. The resulting new materials just might address some of our most vexing problems. To that end, her work has already led to efficient solar cells and powerful batteries (that she hopes one day will be good enough to run her car); a possibly cheaper, greener way of producing plastics; and a potentially better way to peer into deeply buried tumors in the breast and abdomen.

Nature has done an amazing job of making materials that create and feed themselves with readily abundant, non-toxic resources. But it took a long time to get good at this. The first life forms appeared 500 million years ago, and the big explosion of diversity in the Cambrian period took 50 million years. Instead of waiting 50 million years, Belcher is speeding up the evolutionary process by running 1 billion experiments at a time. She starts with a billion viruses, harmless to everything except bacteria that have been genetically altered so they each create slightly different proteins. These viruses are mixed together with whatever elements Belcher has chosen from the periodic table. Out of the billion different proteins the viruses make, roughly 100 will link up with the elements the way she wants them to. Further testing narrows the candidate proteins down to a handful that have promising capabilities. The viruses are the factories producing the material—their genes are programmed to link the organic and inorganic—but they are not present in the final product, so there's no potential risk of viruses running amok. She's found a few candidate viruses that can link methane and oxygen to form ethylene, a building block of plastics, fertilizers, and tires. This ethylene assembly line can take place at room temperature, using natural gas, which is low-polluting and abundant (current production requires lots of energy from high-pollution fossil fuels). It is the ultimate futuristic project and can create many viable solutions to our daily challenges that are non-toxic and eco-friendly.

Spider Silk Electronics

Adding gold to spider silk creates microscopic wires

Scientists chasing the secrets of spider silk have spent years focusing on the material's remarkable mechanical properties. At 2 to 5 micrometers in diameter, it's almost invisibly thin yet incredibly strong for its size, and it can stretch up to four times its original length. And everyone who has ever been in an attic knows how easily it adheres to things. But while most researchers focus on the mechanical aspects, "spider silk, in fact, has an important and promising role in electronics. Scientists studying condensed matter science at Florida State University's National High Magnetic Field Laboratory are exploring what happens when spider silk is coated with iodine, gold, or carbon nanotubes. In all three cases, the silk turned into a more conductive fiber, and gold nanoparticles adhere to spider silk very well.

The research team started with 3.5-μm-wide silk harvested from *Nephila clavipes*, the golden orb-weaving spider. They placed the silk in a vacuum chamber and coated it with gold. The resulting fiber had electrical conductivity from the gold plus flexibility from the silk, and it measured only 1/25th the diameter of a human hair. That allowed the fibers to be used—even without conductive paste—as contacts on tiny organic crystals, which the lab chills to cryogenic temperatures to study their superconductivity. Standard wires made of gold or other metals aren't elastic enough and tend to lose contact with the soft organic crystals as temperatures change. The group found that the gold-coated silk fibers worked as contacts down to the lowest temperature they tested at, about 260 millikelvin. The researchers also coated the silk with carbon nanotubes, creating highly sensitive strain sensors. Those could be used as heart-pulse monitors. Meanwhile, the iodine-doped silk showed changes in conductivity that depended on relative humidity. But the science didn't stop there. They heated the iodine-doped silk in an argon atmosphere to 800°C, creating a pyrolyzed, carbon-coated fiber that turned out to be a *p*-type semiconductor. The researchers then used those fibers to make filaments for incandescent bulbs.

Combining the Florida State research with efforts underway elsewhere to create spider silk artificially might allow engineers to mass-produce fibers with tunable electrical properties.

Eco-Ink

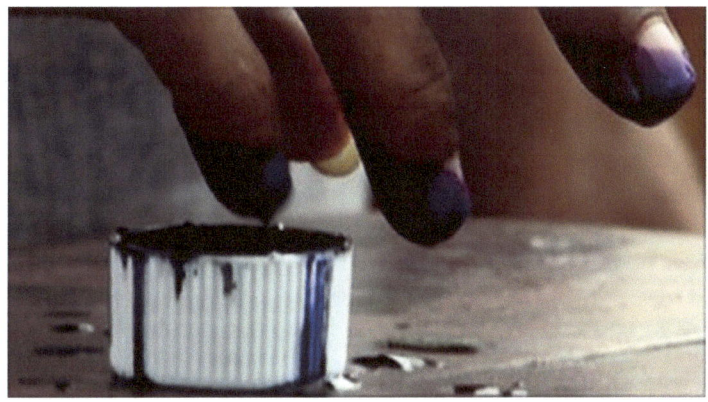

Eco-friendly ink

Although the use of recycled paper has become commonplace, less attention has been paid to the inks used by the printing industry. This is starting to change, with the creation of a new generation of inks that are said to be kinder to the environment. One company at the forefront of these developments is EnNatura. This Delhi-based startup makes a biodegradable ink that, the company claims, also has other benefits, such as making the process of cleaning printing machinery both easier and more environmentally friendly.

The conventional printing process makes use of a lot of petroleum-based chemicals and solvents. These solvents are responsible for photochemical smog formation and ozone depletion, and are also hazardous to the printers. EnNatura's biggest challenge was to create an ink that would not only be eco-friendly, but would also have the vivid colors of conventional ink. They solved this by developing their own resin. Other manufacturers are just picking up resin off the shelf and making ink out of it, which is very damaging to the environment.

Self-Cleaning Fabric

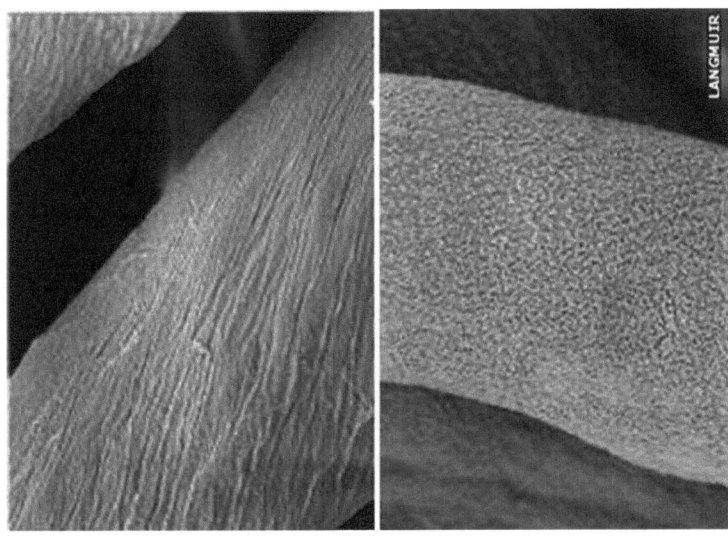

Uncoated fibers   Silica coated fibers

Super-hydrophobic surfaces have fascinated scientists for years; they are behind the lotus plant's self-cleaning leaves and the gecko's super-dry and thus super-sticky feet. These surfaces are practically impossible to wet—water beads on them, and dirt and particulates do not stick to them, leading to the self-cleaning property. Chemists looking for the next big thing in clothing coatings have tried several tricks in recent years to create a coating with similar properties in the laboratory, but have failed to produce a durable coating. However, recently a group of chemists from the Australian Future Fibres Research and Innovation Center at Deakin University devised a better method of coating fabrics with a water-repellent, "self-cleaning" coating. The trick was to engineer a multi-layered coating whose layers, when struck with UV light, bond more firmly to each other and to cotton. The technique may also be put to use in medical antibacterial coatings.

The new work hinges on what is known as layer-by-layer self-assembly—basically dipping a fabric into a solution over and over again to deposit multiple layers on it. The team made their solution with tiny particles of silica, the same material as sand. Crucially, they added a few chemical steps to coat the particles with long chemical tails ending in what are known as azido groups. The team found that layered coatings of the azido-treated nanoparticles were, like many previous attempts, not very durable; a stain-free shirt with such a coating would not survive many washes. But when baked under a source of UV light, the tails on the particles were made to interlink with one another, forming a far tougher structure within and across the layers. At this stage, the technique is only suited to small samples.

SECTION 4

COOL TECHNOLOGY

Touch-Less Surgery

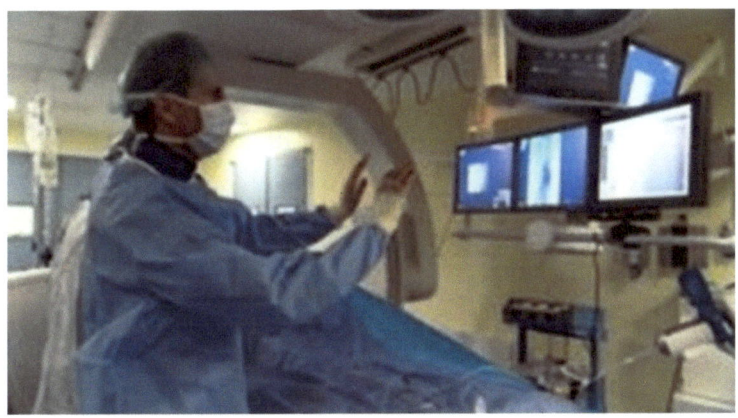

Trial of "touch-less" gaming technology in surgery

Surgeons at St. Thomas' hospital in London are conducting trials on gesture-based gaming technology to access and manipulate images. The system will be familiar to anyone who has used Kinect interactive games at home. This has been adapted to respond to a surgeon's voice commands and arm-movements during operations.

The initial trial at St. Thomas' hospital was in vascular surgery, in this case inserting a graft to repair a damaged aorta, the main blood vessel running through the body. Standing straight, arms raised like a conductor, the surgeon issues commands to a Kinect sensor perched beneath a monitor displaying a 3D image of the patient's damaged aorta. With hand gestures he can pan across, zoom in and out, and rotate images. He can then lock the image and make markers to help ensure the graft is in exactly the right place. Until recently surgeons were shouting out across the operating theatre to tell someone to go up, down, left, or right. But with the Kinect they are able to get the position that they want quickly, without having to handle non-sterile objects, such as a keyboard or mouse, during the procedure. This is one of the first trials of its kind in the world. Some of the features—such as the voice control and gestures tailored to vascular surgery—are unique. The refinements from gaming technology to complex surgery have been developed by Microsoft Research, with support from Lancaster University. This trial will soon be extended to other centers and other types of surgery. These advances in image manipulation, which is an integral part of the minimally invasive surgery that is done nowadays, are inevitably going to become more refined and easily available. This is a great example of a successful interdisciplinary research project, combining the technical skills of computer scientists with social scientific and medical expertise to ensure that the new technology resonates with the way in which surgeons actually do their work.

Ultrasound for Fractures

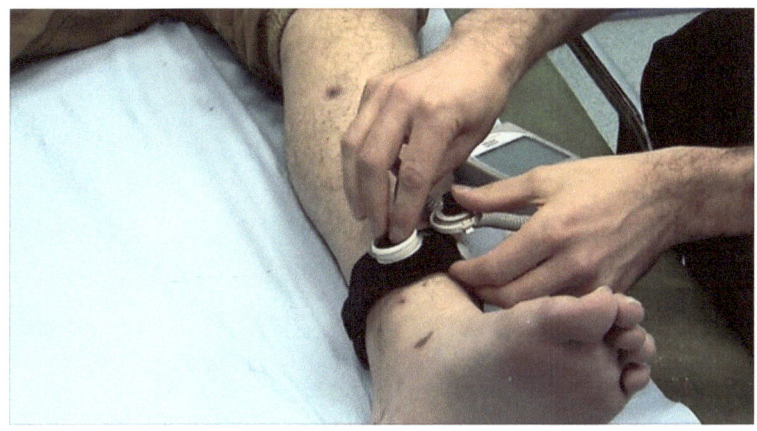

Ultrasound speeds up recovery for patients with broken bones

Doctors in the Scottish city that pioneered the use of ultrasound to scan the body are now using ultrasound to heal broken bones. Orthopedic surgeon Angus MacLean has been using the technology at Glasgow's Royal Infirmary fracture clinic. The evidence suggests that ultrasound speeds up the process by about 40 percent. The technology is similar to that used on pregnant women. There's strong evidence that ultrasound slightly vibrates the cells, which then stimulates healing and regeneration in the bone. Ultrasound waves are used at a slightly different frequency and a slightly different pulse. Research suggests this encourages cells to remove bacteria, stimulates the production of new bone cells, and encourages those cells to mature more rapidly. Because of the costs involved—around $2,000 per patient—ultrasound is only being used on complex fractures. It is expected that the cost of using ultrasound to treat fractures will reduce over time, making it a cheap way to speed up the healing of common fractures as well as complex ones.

Ultrasound was first developed as a diagnostic tool in Glasgow in the 1950s. A team of specialists, led by Professor Ian Donald, produced the first images of the body using a technology adapted from sonar at Glasgow's Western Infirmary. It has become one of the most common medical technologies in the world. But it is only now, 50 years later, that its potential for aiding the healing process is being unlocked.

Self-Driving Car

Google gets Nevada driving license for self-drive car

Driverless cars will soon be a reality on the roads of Nevada after the state approved America's first self-driven vehicle license. The first to hit the highway will be a Toyota Prius modified by the search firm Google, which is leading the way in driverless car technology. Its first drive included a spin down Las Vegas' famous strip. Other car companies are also seeking self-driven car licenses in Nevada. The car uses video cameras mounted on the roof, radar sensors, and a laser range finder to "see" other traffic. Engineers at Google have previously tested the car on the streets of California, including test runs on San Francisco's Golden Gate Bridge. For those tests, the car remained manned at all times by a trained driver ready to take control if the software failed. The car has covered 140,000 miles with no accidents, other than a bump at a traffic light from a car behind it. Nevada changed its laws to allow self-driven cars in March. The long-term plan is to license members of the public to drive such cars. To make it recognizable, Google's car has been issued with a red license plate. The plate features an infinity sign next to the number 001. Other states, including California, are planning similar changes. As we know, the vast majority of vehicle accidents are due to human error. Through the use of computers, sensors, and other systems, an autonomous vehicle is capable of analyzing the driving environment more quickly and operating the vehicle more safely.

Self-Driving Convoy

Volvo's self-drive 'convoy' hits the Spanish motorway

In the first public test of such vehicles, a convoy of self-driven cars has completed a 200 kilometer (125 mile) journey on a Spanish motorway. The cars were wirelessly linked to each other and "mimicked" a lead vehicle, driven by a professional driver. The so-called road train has been developed by Volvo. The firm is confident that the vehicles will be widely available in future. The project aims to herald a new age of relaxed driving. Drivers can now work on their laptops, read a book or sit back and enjoy a relaxed lunch while driving. The road train test was carried out as part of a European Commission research project known as Sartre—Safe Road Trains for the Environment. The convoy comprised three cars and one truck. The cars are fitted with special features such as cameras, radar, and laser sensors, allowing each vehicle to monitor the lead vehicle and also other vehicles in their immediate vicinity. Using wireless communication, the vehicles in the platoon "mimic" the lead vehicle using autonomous control—accelerating, braking, and turning in exactly the same way as the leader. The vehicles drove at 85 kph (52 mph) with the gap between each vehicle just 6 meters (19 feet).

People think that autonomous driving is science fiction, but the fact is that the technology is already here. Apart from the software developed as part of the project, it is only the wireless network installed between the cars that sets them apart from other cars available in showrooms today. Everything should function without any infrastructure changes to the roads or expensive additional components in the cars. The three-year Sartre project has been under way since 2009. Other partners include UK car technology firm Ricardo UK, Tecnalia Research & Innovation of Spain, Institut fur Kraftfahrzeuge Aachen (IKA) of Germany, and the Technical Research Institute of Sweden. The eventual aim of the project is to have several cars "slaved" to a lead vehicle and travelling at high speed along specific routes on motorways.

Robotic Fish

Robotic fish to patrol for pollution in harbors

In the shallow waters of Gijon harbor, in northern Spain, a large, yellow fish cuts through the waves. But this swimmer stands apart from the marine life that usually inhabits this port: there's no flesh and blood here, just carbon fiber and metal. This is robo-fish, scientists' latest weapon in the war against pollution. This sea-faring machine works autonomously to hunt down contamination in the water, feeding this information back to the shore. In Spain, several robo-fish are undergoing their first trials to see if they make the grade as future marine police. The port at Gijon, in northern Spain, is being used as a testing environment. The idea is to have real-time monitoring of pollution, so that if someone is dumping chemicals or something is leaking, it is easy to find out what is causing the problem and to put a stop to it. The company is part of the Shoal consortium, a European Commission-funded group from academia and business that has developed these underwater robots. Currently, they take samples in harbors about once a month. And in that time, a ship could come into the harbor, leak some chemicals somewhere, and continue all the way up the coastline. The idea is that robot fish will be in the harbor all the time, constantly checking for pollution.

The fish, which measure about 1.5 meters long, may be a little larger than their real-life counterparts, but their movements closely mimic them. They swim just like fish; they are really quite agile and can change direction quickly, even in shallow water. Traditional robots use propellers or thrusters for propulsion, and the robo-fish uses its fin to propel through the water. The fin is a useful tool in shallow waters, especially where there is a lot of debris. It can work in environments that are very weedy and would usually snag propellers. The fish use micro-electrode arrays to sense contaminants. In their current form, they can detect phenols and heavy metals such as copper and lead, as well as monitor oxygen levels and salinity. But its design is flexible, and one chemical sensor unit can be replaced with different ones for something else, such as sulphates or phosphates, depending on the environment that needs to be monitored.

Once they've sniffed out a problem, the fish use artificial intelligence to hunt down the source of pollution. They can work alone or in a team, communicating with each other using acoustic signals, and they can continuously report back to the port. The trials at Gijon have been designed to put all of this technology to the test to finalize the design of the robots. In the future, it will develop into multitasking robo-fishes—robots that can do search and rescue and monitoring for underwater divers at the same time as tracking pollution. Prototypes currently cost about $40,000 each; costs will drop once more are produced. But battery life is also an obstacle. Currently, the fish need to be recharged about every eight hours.

Mine Kafon

A cheap, wind-blown mine clearance device

Massoud Hassani, an Afghani, designed and built, by hand, a wind-powered ball that is heavy enough to trip mines as it rolls across the ground. Each $50 device looks like artwork inspired by a starburst. In the middle of the Kafon is a 17 kilogram (37 pound) iron casing surrounded by dozens of radiating bamboo legs that each has a round plastic "foot" at their tip. Inside the ball is a GPS unit, which maps where it has been, and in theory, that area is cleared of mines. Around the iron ball is a suspension mechanism, which allows the entire Kafon to roll over bumps, holes, and other obstacles. In all, it weighs a little more than 80 kilograms (175 pounds). The idea is that it is light enough to be pushed by the wind, but heavy enough to trip mines. Hassani thinks that humanitarian organizations could take Kafons with them into areas suspected of being mined, and then let the wind do the dangerous work.

More countries deployed anti-personnel mines last year than in any year since 2004, according to an international survey of landmines. The global statistics on land mines and their effects make sobering reading. According to the United Nations, up to 110 million mines have been laid across more than 70 countries since the 1960s and that between 15,000 and 20,000 people die each year because of them. Many of the victims are civilians—children, women and the elderly—not soldiers. Thousands more are maimed. Moreover, mines are cheap. The UN estimates that some cost as little as $3 to make and lay in the ground. Yet, removing them can cost more than 50 times that amount. And the removal is not without human cost either. The UN says that one mine clearance specialist is killed, and two injured, for every 5,000 mines cleared. One of the worst-affected countries is Afghanistan, with an estimated 10 million land mines contaminating more than 200 square miles. Normally dogs and mine detectors are used to find mines. But that takes a lot of time and costs a lot of money, because it requires digging out every metal part that is detected. No one knows whether it is actually a mine or not until it is excavated. Devices like the Kafon can speed up this process and take some of the danger out of the equation.

The device needs some improvement before it can be used effectively for de-mining activities. The idea that the Kafon can hit a mine, survive the blast, and then continue to roll on and detonate other mines and eventually clear an area may not work. It is more likely that 100 grams of explosive (the average for anti-personnel mines) will stop the Kafon dead it in its tracks right in the middle of a probable mine field, which would make retrieval and repair very dangerous. There is also the possibility that a rolling Kafon would activate trip wires on fragmentation mines. If the mine explodes into fragments, it causes more problems, because in subsequent sweeps every fragment will be detected as a potential mine. In other words, more metal pieces mean more work for the human de-miners who have to prove 98 percent clearance. There are also a couple of other obvious drawbacks. First, the wind must be blowing for the Kafon to move. And second, it will only work on fairly flat and open terrain, such as a desert. It isn't designed to move in jungles and other terrains. However, the Kafon was recently selected as a finalist for the 2012 Design of the Year award at the Design Museum in London, and Hassani is working hard to improve the device.

Voice-Controlled Television

Samsung's 'future-proof' voice-controlled television

Samsung has unveiled a "smart" internet-connected television that has the ability to have its hardware upgraded every year. The device has a slot which allows the user to add new kits that boost processing performance and add new features. The innovation may help reassure shoppers concerned about their screen becoming outdated. In addition to the television's "smart evolution capability," Samsung also added gesture, voice, and face recognition features to the ES8000 model. A built-in camera allows users to browse the internet with a wave of their hand and to change channels by speaking in one of the more than 20 languages that the set can understand. A facial recognition facility also allows the set to recognize users, pulling up the relevant selection of their favorite apps. The device is the latest in a run of so-called Smart TVs launched by the firm since 2008.

Connected televisions with built-in processors were one of the hottest trends at this year's Technology trade show. It is expected that about half of all shipped TVs would have internet capabilities in 2012. But the figure was 12 percent of all units shipped in 2010.

Virtual Grocery

Shopping by phone at South Korea's virtual grocery

Online shopping is nothing new, especially in plugged-in South Korea. But one company says it's going further. Homeplus is testing out a virtual supermarket in a public place. At Seolleung underground station in Seoul, there's a row of brightly lit billboards along the platform, with hundreds of pictures of food and drink—everything from fruit and milk to instant noodles and pet food.

Earlier this year, the *Wall Street Journal* reported that the number of Smartphone subscribers in South Korea had passed 10 million, up from just a few hundred thousand in 2009. Young Koreans increasingly rely on smart phones to take care of many of their daily tasks, and the virtual store allows them to save time. All one needs to do is to hold the smart phone over the black-and-white QR—the Quick Response code—under the picture of the item. There's a beep, and the picture of the item appears on the phone screen. After of the user selects an item, the app asks them to enter when and where the product should be delivered. If orders are placed before 1:00 p.m., the company pledges to deliver the groceries the same evening.

Homeplus plans to put virtual stores in other underground railway stations, especially those close to the city's universities. And the company wants to introduce them in other countries, too.

Mirage Table

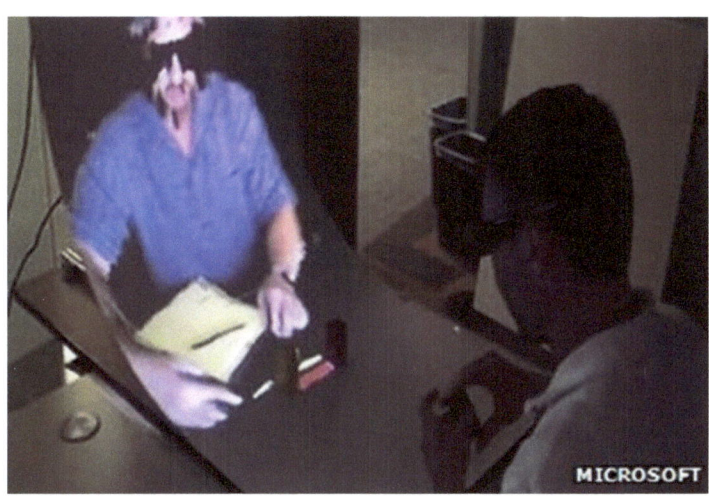

Augmented reality device creates an illusion of reality!

Microsoft has developed an augmented reality system that allows users at different locations to work together on tabletop activities, sharing objects which they both can handle. It could "fool" the eye to suggest both parties were using a seamless 3D shared task space. The Mirage Table uses a 3D video projector to beam images onto a sheet of curved white plastic placed in front of the user. At each end, one of Microsoft's Kinect depth camera sensors is used to track the direction of each person's gaze as well as the participant sitting behind them, and to capture the shape and appearance of objects placed on the surface. Users are also required to wear shutter glasses in order to see the projected image in three dimensions. Two computers linked by a network connection are required to power the experience. The researchers were motivated by a simple idea: Can a user interact with 3D digital objects alongside real objects in the same physically realistic way and without wearing any additional trackers, gloves, or gear?

This type of experience marks a significant improvement on current video conferencing technologies. In this system, the user can hold a virtual object, move it, or knock it down, since all virtual and real objects participate in a real-world physics simulation. The unique benefit of this setup is that two users share not only the 3D image of each other, but also the tabletop task space in front of them. However, the system needs improvements before it can be used. At present the Kinect device only captures the front face of objects, leaving gaps and imperfect texturing. The setup also only allows users to scoop or catch objects from below in order to hold them in their hands.

Ultra-Thin Glass

Ultra-thin glass can 'wrap' around devices

A new type of flexible ultra-thin glass has been unveiled by the company that developed Gorilla Glass. Dubbed Willow Glass, the product can be "wrapped" around a device, such as a smart phone, and could also be used for displays that are not flat. But until such "conformable" screens appear on the market, the glass could be used for mobile devices that are constantly becoming slimmer. The prototype demonstrated in Boston was as thin as a sheet of paper, and the company said that it can be made to be just 0.05 millimeters thick— thinner than the current 0.2 millimeter or 0.5 millimeter displays. The firm has already started supplying customers, who are developing new display and touch technology, with samples of the product.

The material used to make Willow Glass is the result of the firm's glassmaking process called Fusion. The technique requires melting the ingredients at 500^0C, then producing a continuous sheet that can be rolled out in a mechanism similar to a traditional printing press. This roll-to-roll method is much easier and faster for mass production than the sheet-to-sheet process normally used to make super-thin glass. In the future,

Willow Glass may replace the already widely used Gorilla Glass, which is found on many smart phones and tablets. The first generation of Gorilla Glass, launched in 2007, has so far been used on more than 575 products by 33 manufacturers, covering more than half a billion devices worldwide. It was first spotted by the Apple founder Steve Jobs, who contacted Corning when the firm was developing the screen for its first iPhone in 2006. Willow Glass is not the first attempt to produce a futuristic flexible display. During the past few years, scientists around the world have been working with a material called graphene, a super-conductive form of carbon made from single-atom-thick sheets, first produced in 2004.

Virtual Changing Room

Virtual changing rooms make shopping a wonderful experience!

Sometimes trying on clothes can be fun; it feels great when you find those perfect jeans or a dress that you never want to take off. Other times, it can be a bit of a confidence crusher and can eat into valuable shopping time. Now shoppers can 'try on' garments in a virtual changing room without removing a single item of clothing. Using a 58-inch plasma screen and a 3D depth-sensing camera, customers can superimpose tops, dresses, and more onto a live picture of their body. The device uses technology similar to that of the Xbox Kinect to sense movement and calculate your size. To select a different size, you simply wave your hand, a much more dignified alternative to calling out for an attendant while half-dressed behind a curtain. Once you've found the perfect fit, you can make your way to the store to pick it up. This is a secret weapon against the rise of internet shopping; customers just can't get this type of theatre and experience online. This puts the excitement back into shopping.

Google Fiber

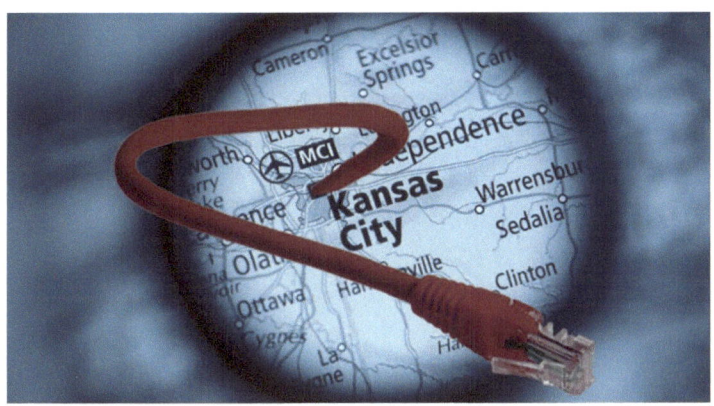

Can ultra-fast internet change a city?

Google, the California internet giant, has begun installing fiber optic cable that will give Kansas City residents download speeds of up to 1Gbps—about 100 times faster than the broadband internet service currently available to most Americans. Although the seeds of the internet germinated in US Department of Defense laboratories and many of the most innovative internet companies are based in the United States, Americans have far slower internet than residents of many other industrialized nations. The average broadband internet speed across the US is 12.84 Mbps, according to Netindex.com. That makes the US 31st in the world.

The high speed will enable small businesses and home-office workers to have high-definition video conferencing without the hiccups, lag-time, and buffering slogs frequently suffered with cable or DSL broadband. It will allow greater use of cloud computing by small businesses by, for example, allowing them to keep customer databases and accounting systems online instead of in costly local servers. Having affordable super-fast internet in the home will enable faster and more efficient telecommuting, which could take cars off the roads. It will be easier for doctors and hospitals to transmit data-heavy medical images, such as MRI scans. Businesses or local governments could install "dumb terminals"—computers with little more than a screen, keyboard, mouse, and internet connection—across the city. Communities could establish shared music, film, and e-book libraries. High definition—even holographic—video conferencing could enable greater participation in local government; "town hall in the home" is one catchphrase. Public safety could be improved by higher definition CCTV and video emergency calling.

When electric power first became widely available in homes, it was a more convenient, somewhat novel alternative to the oil lamp. At that time, it would have taken an incredible visionary to predict what kind of an impact electric power would have on business and, ultimately, quality of life. By the same token, it is difficult to predict what the overall impact of Google fiber will be.

Everyday Objects

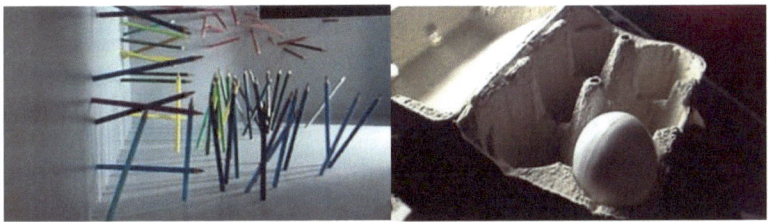

The mundane invention!   Eggcellent discovery!

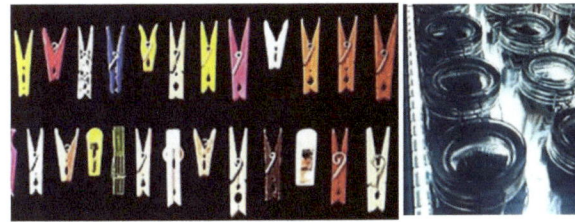

Got it pegged?   The classic Kilner jar!

Most of us take for granted everyday objects like a paper clip, a clothes peg, a teabag, the egg box, and the zip fastener—and we are not curious to explore the stories behind their invention. Who knows, for instance, that the rawlplug that millions of us use in DIY jobs every day was invented in 1910 by an engineer called John Joseph Rawlings? He was contracted to install electrical fittings at the British Museum with the stipulation that the walls should be damaged as little as possible. He came up with a plug made out of jute fibers that had been saturated with glue. Fifty years later a German inventor developed a plastic plug that used the same principle of "grip by expansion." Then there is the clothes hanger. It looks very simple, but 189 patents were issued for various models between 1900 and 1906. One was for a wire hanger invented when Albert Parkhouse arrived at work on a cold winter's day to find all the coat pegs taken—so on the spot he bent a piece of wire into a hanger. The coffee filter came about in n 1908, when German housewife Melitta Bentz invented an effective coffee filter by lining a perforated metal beaker with blotting paper. It was 30 years before this evolved into the form it has today, but despite the arrival of all sorts of smart and expensive machines, the simple paper filter is still the easiest route to a good cup of coffee in millions of homes around the world. More recent products that already look destined for longevity include the sticky notes plastered on computer screens, notice-boards, and documents just about everywhere. This all began with a failure in the late 1960s, when an American scientist at 3M's research lab, Dr. Spencer Silver, was trying to develop an extra-strong adhesive. Instead, he developed weak glue that allowed things to be joined and taken apart easily. A decade later, Silver's colleague Arthur Fry, irritated by paper bookmarks which kept falling out of his hymnbook during choir rehearsals, decided to coat them with the weak glue. The Post-It note was born. But, one must understand that the supposedly mundane objects are often remarkable because they have longevity. Decades from now we will still find a rubber band more useful than an iPhone (which by then will be a dinosaur because it is not simple enough). From the paper clip to the Post-It note, simple and very cheap products have proved that they can have lasting appeal, and make some of their inventors very wealthy.

www.ingramcontent.com/pod-product-compliance
Lightning Source LLC
Chambersburg PA
CBHW041103180526
45172CB00001B/87